Acorn

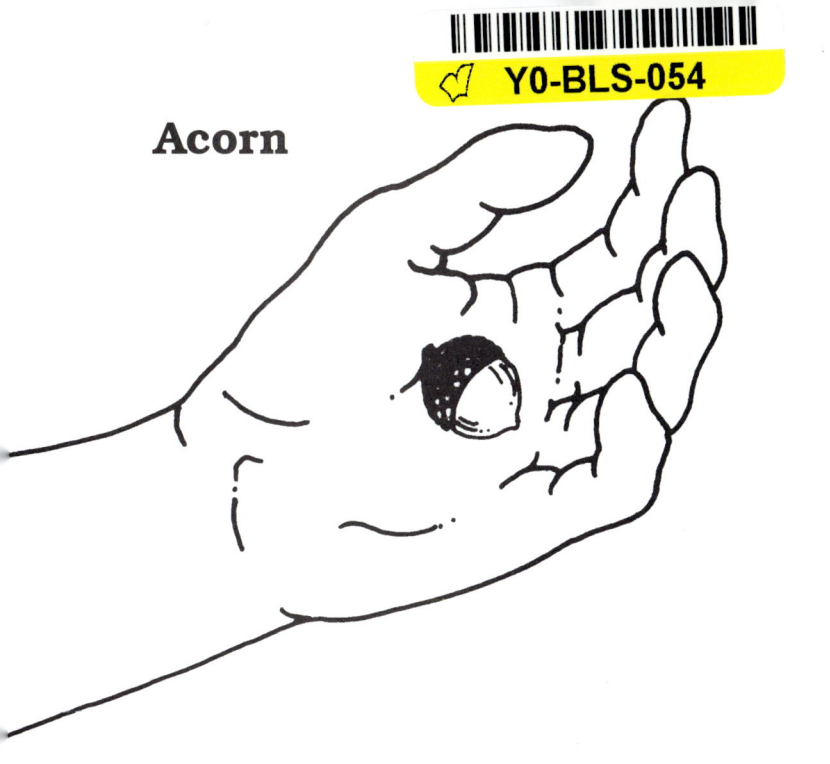

This is just a tiny acorn. It will not be an acorn for long. What do you think it will grow into?

The acorn will grow into
a tree. It will stand, silent
and strong. It will be home
to birds and spiders.

Iris bulbs

These bulbs will be planted deep in the dirt. You won't even be able to see them! They'll be hidden in the earth.

Rain and sun help the bulbs grow. Tiny shoots will pop up. In the spring, you will have an iris!

Cactus

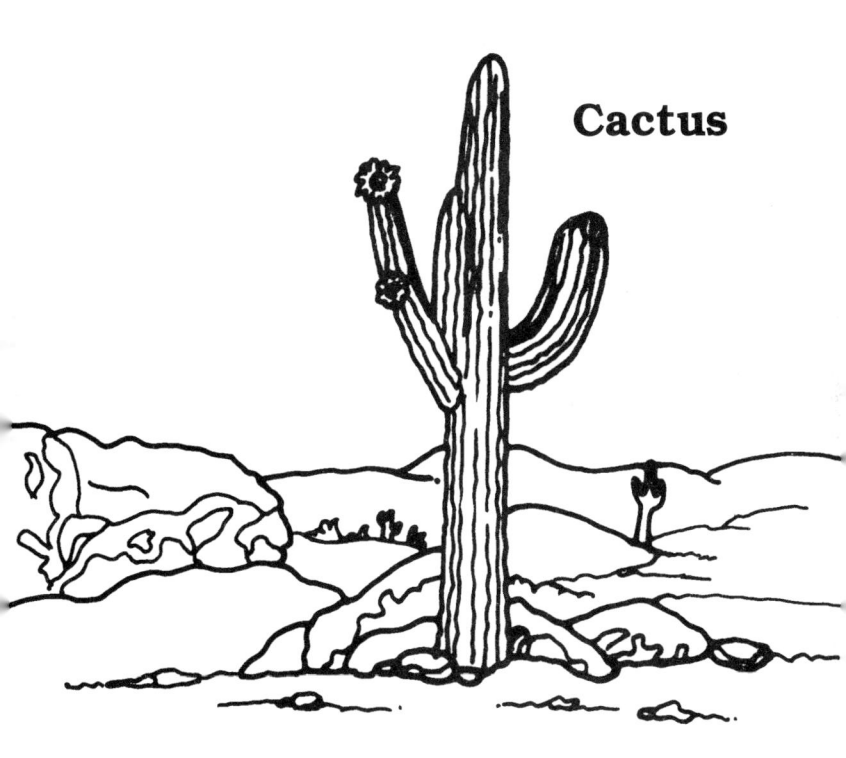

Some plants are able to grow in sand. This cactus does not need much rain. Even a little rain will do!

Plants like to grow in dirt.
Even these plants grow
in earth. The earth is at the
bottom of the pond.

Venus's-flytrap

Plants get food from the earth. This plant traps its own food. It gets food from bugs!

Bamboo

In China, bamboo plants grow high. This baby's mother has bamboo for dinner!

Understanding the Story

Questions are to be read aloud by a teacher or parent.

1. What is the book about?

2. What do plants need to grow?

3. What kind of plants do pandas eat?

4. Which plant do you like best? Why?

Answers: 1. plants 2. rain, sun, earth 3. bamboo 4. Answers will vary.

Saxon Publishers, Inc.
Editorial: Barbara Place, Julie Webster, Grey Allman, Elisha Mayer
Production: Angela Johnson, Carrie Brown, Cristi Henderson

Brown Publishing Network, Inc.
Editorial: Marie Brown, Gale Clifford, Maryann Dobeck
Art/Design: Trelawney Goodell, Camille Venti, Andrea Golden
Production: Joseph Hinckley

© Saxon Publishers, Inc., and Lorna Simmons

All rights reserved. No part of the material protected by this copyright may be reproduced or utilized in any form or by any means, in whole or in part, without permission in writing from the copyright owner. Requests for permission should be mailed to: Copyright Permissions, Harcourt Achieve Inc., P.O. Box 27010, Austin, Texas 78755.

Published by Harcourt Achieve Inc.

Saxon is a trademark of Harcourt Achieve Inc.

Printed in the United States of America
ISBN: 1-56577-991-6

5 6 7 8 546 12 11 10 09 08

Saxon Phonics and Spelling 1

Phonetic Concepts Practiced

ā´|cv (acorn)
ĭ´|cv (tiny)
ē´|cv (even)

ISBN 1-56577-991-6

Grade 1, Decodable Reader 29
First used in Lesson 82

A Hobgoblin Saves the *Atlantic*

written by Cynthia Benjamin
illustrated by Kate Flanagan

THIS BOOK IS THE PROPERTY OF:

STATE_____
PROVINCE_____
COUNTY_____
PARISH_____
SCHOOL DISTRICT_____
OTHER_____

Book No. _____
Enter information in spaces to the left as instructed

ISSUED TO	Year Used	CONDITION ISSUED	RETURNED

PUPILS to whom this textbook is issued must not write on any page or mark any part of it in any way, consumable textbooks excepted.

1. Teachers should see that the pupil's name is clearly written in ink in the spaces above in every book issued.
2. The following terms should be used in recording the condition of the book: New; Good; Fair; Poor; Bad.